A SUMMARY, REVIEW &

ANALYSIS OF

ELIZABETH KOLBERT'S

THE SIXTH

EXTINCTION

BY

SAVE TIME SUMMARIES

Note to Readers: We encourage you to first order a copy of Elizabeth Kolbert's full book, _The Sixth Extinction_ before you read this unofficial Book Summary & Review. Most readers use this guide by first reading a chapter from the full copy, and then reading the corresponding section from this Book Summary & Review. Others prefer to read the entire book from cover-to-cover, and then review using this review and analysis.

A SUMMARY, REVIEW & ANALYSIS OF ELIZABETH KOLBERT'S

Other Amazon Kindle Ebooks from *Save Time Summaries:*

Summary of Malala Yousafzai's *I am Malala*

Summary of Malcolm Gladwell's *David & Goliath*

Summary of Dr. David Perlmutter's *Grain Brain*

Summary of Robert Lustig's *Fat Chance -- Battling Sugar, Obesity & Disease*

Summary of Stephen Covey's *The 7 Habits of Highly Effective People*

TABLE OF CONTENTS

OVERVIEW

The idea of extinction is a fairly new one, never considered as even a possibility until the end of the eighteenth century, despite the discovery of mammoth and mastodon bones in North America, Europe, and Russia decades earlier. Naturalist Georges Cuvier was the first person to study these fossils and declare their existence as proof of "lost" species.

Over the course of the past half-billion years, earth has undergone five mass extinction events, which scientists refer to them as the Big Five. The cause of these events varies: glaciation, global warming, asteroid collision. The result, however, was the same: conditions on earth changed too rapidly for most organisms to adapt, and those species perished. Scientists believe the earth is in the middle of the sixth big event, the Sixth Extinction. This time, the cause is humans.

Around the world, species are disappearing at a rapid rate. This is not a new phenomenon; species have been vulnerable to the influence of humans since modern human beings made their first appearance tens of thousands of years ago. The reasons behind these extinctions and near-extinctions vary: loss of habitat, loss of resources, hunting and poaching, disease. However, the illustrator of these catastrophes has been the same each time: humankind.

In *The Sixth Extinction*, Elizabeth Kolbert lays out the many ways humans impact their world, a world they share with millions of other life forms. Kolbert's narrative of earth's history is easy to understand and delivered with a light touch and occasional flashes of humor that help leaven the overall mood. Mass extinction is dark subject matter for a book, but Kolbert makes it both interesting and entertaining. Her "in the field" time with scientists of many disciplines is evident on every page.

This book will likely be an eye-opener for many readers. The impact humans have had on the planet, beginning in their earliest days when they began migrating to every corner of the globe, is astonishing simply because it has been so far-reaching. The powerful impact of seemingly innocuous changes to their environment may come as a shock, such as the destruction that may be wrought by a single foreign seed. Of particular import are the changes and destruction that can be laid at humanity's feet today. After all, focusing on the happenings of thousands of years ago won't change them. In fact, it's not likely that humanity's current course can be altered. The interesting and sobering thought is that, long after humanity has been reduced to a layer of rock the width of a sheet of paper, its legacy will continue through the earth's new landscape.

PROLOGUE

Summary

Humans began modestly, a small group of small beings (relative to surrounding creatures), confined to a small area. Prodigious breeding, cognitive intelligence, and a highly adaptive nature allowed them to expand to all areas of the globe. In a short time, universally speaking, their population increased eight-fold. The impact humans had on their environment was immense, from decimation of resources to species extinction.

Over the lifetime of the planet, there have been five cataclysmic events, called the Big Five, which caused radical change, as well as numerous other catastrophic events that were less impactful. Earth is on the precipice of the sixth event, the Sixth Extinction.

Key Take-Aways

- Humankind spread rapidly across the planet, leaving a trail of destruction in its wake.

- Earth is poised to experience its sixth radical, cataclysmic event.

CHAPTER 1

Summary

The golden frog was a fixture in El Valle de Anton, Panama, found everywhere. Highly toxic, their golden coloring makes them stand out vividly against their environment. However, where they were once prolific, they are now rapidly disappearing, faster than biologists' efforts to combat their extinction.

The first mass extinction happened 450 million years ago. Two hundred million years later, the second one struck, wiping out nearly all life on the planet. The Cretaceous period, the most recent mass extinction, destroyed the dinosaurs, plesiosaurs, mosasaurs, ammonites, and pterosaurs. Herpetologists believe a similarly devastating event is currently striking amphibians.

(Photo courtesy of Commons.wikimedia.org)

In the attempt to stop their extinction, the El Valle Amphibian Conservation Center (EVACC) houses a number of at-risk amphibians, including the golden frog. Upon realizing the danger, biologists began collecting frogs as quickly as they could, but some species went extinct before they could be brought to EVACC.

The sudden disappearance of frogs from areas where they used to run rampant is mysterious, because they're vanishing even from "pristine" areas, not only those that have been disrupted. Additionally, frogs with vast populations are disappearing just as quickly as those with limited numbers. Clues to the frogs' disappearance came from The National Zoo in Washington, D.C., which also experienced mysterious deaths in its frog population. Scientists there discovered a new form of chytrid fungus on the frogs' skin, which interferes with their ability to absorb electrolytes. They die within a few weeks of exposure to the fungus. The fungus thrives even without amphibians, so EVACC's ultimate goal, sending the frogs back to their natural habitat, seems impossible. The fungus has spread and is now found across the globe, most likely due to humans ferrying infected frogs to foreign locations.

In normal geological epochs, extinction is rare and is generally called "background extinction," which varies from one life form to the next. As a rough average, mammalian species become extinct once every 700 years. Mass extinctions happen in a relatively short amount of time and involve global, "substantial biodiversity losses." There have been many mass extinctions, all of which were less catastrophic than the Big Five. Amphibian extinction is rarer than mammal extinction, yet currently they are in the most endangered class of animals. Similar losses are occurring in other groups across the globe.

Key Take-Aways

- Many species of amphibians, and particularly frogs, are threatened with mass extinction due to a sudden change in their environment: the appearance of a new breed of fungus.

- Mass extinction is historically rare, but it has experienced an alarming increase. Most incidents of mass extinction can be traced back to humans.

CHAPTER 2

Summary

Early scientists did not consider the history of animals, and Carl Linnaeus' system of binomial nomenclature included only living creatures. Despite discoveries of strange bones and fossils, extinction was never considered until the discovery of an American mastodon in 1739 by the second Baron de Longueuil, Charles le Moyne, and the work of naturalist Jean-Léopold-Nicolas-Frédéric Cuvier in the late 18th century. Longueuil brought the bones, including femur, tusk, and teeth, back to France, where scientists were flummoxed by its teeth. Numerous theories arose, but it was unthinkable that the bones were from an extinct creature. More mastodon bones were uncovered, in America, Russia, and England, and still no one considered extinction.

In 1795, Cuvier secured a position at Paris's Museum of Natural History and studied the mysterious bones in his spare time. He presented his theories, including that the collection of mammoth bones from the different locations came from disparate species, as evidenced by their teeth. He also concluded that they were "lost" species. He made similar studies of the bones of other animals, and came to the same conclusion: these species are extinct. He went further, claiming there must be other extinct species, and there was evidently "a world previous to ours." Cuvier's work and theories were revolutionary, especially considering the lack of evidence and popular opinion.

Cuvier established extinction as fact, but his lost world theory remained theory. Therefore, he became determined to find the fossils that would prove it. He gathered specimens from collectors and naturalists: sometimes bones and sometimes detailed drawings. Four years after his first presentation, he'd identified 23 extinct species, insisting there must be many more. In 1812, he published a compendium of his findings, which included 49 extinct species, and he and his theories grew in popularity. The wealthy began collecting fossils, paying "fossilists" (people who hunted and sold fossils) for their finds. This unearthed a host of newly discovered, extinct species.

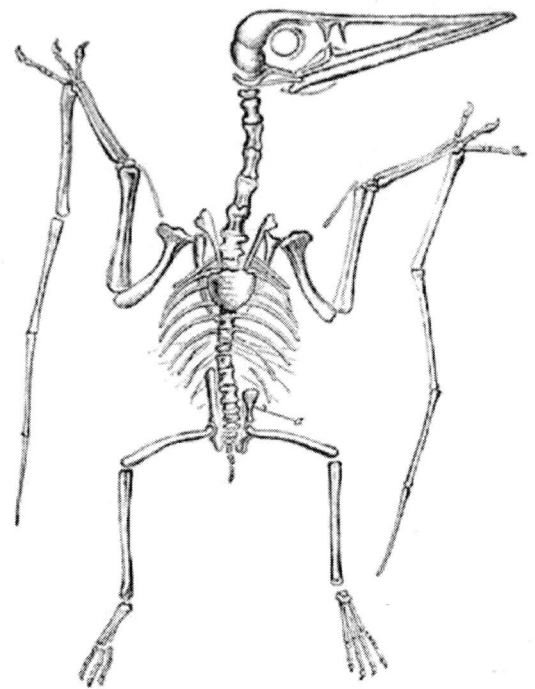

The skeleton of a Pterodactyle. The creature is lying on its back, with the head bent to the left side. *a* indicates the left pubic bone; the haunch-bone, or ilium, being shown on the opposite side. (*After Von Meyer.*)

(Image courtesy of Wikipedia.org)

Studying the layers of the earth, stratigraphy, was a new discipline, but it gave Cuvier another startling insight. Species whose remains were uncovered nearer earth's surface belonged to still-existent orders. Digging further revealed mammals with no obvious tie to current species, and even further down mammals disappeared completely, proving the world was once inhabited by giant reptiles. However, Cuvier did not believe in evolution. He could not account for how new species came to be; his only interest was in their extinction, which he surmised was the result of catastrophe. He found evidence of this theory in stratigraphy. Though now disproved (he discounted the idea of slow change for the earth, as he did for evolution), some of his claims were accurate: cataclysmic events did happen, they just weren't the only agents of change.

Key Take-Aways

- Until Cuvier, no one believed species went extinct.

- Cuvier posited that cataclysmic events caused enormous changes in the earth and its life forms.

CHAPTER 3

Summary

In line with Cuvier's theories, William Whewell coined the term "catastrophist" in 1832, considering himself and most scientists catastrophists. An exception was Charles Lyell, whom Whewell referred to as a "uniformitarian." Lyell disagreed with Cuvier's theories that the earth changed in response to cataclysmic events. He saw no reason to suppose change happened differently than it does now: the result of gradual processes millennia in the making. His views on extinction and speciation were also opposed to those of Cuvier. He didn't believe that life forms progressed from reptiles to mammals, but that all creatures existed throughout all eras, and extinct species could reappear. Lyell's published theories were a best seller of its day.

Charles Darwin was a fan of Lyell. He discovered Lyell's work during his time on the HMS *Beagle*, and it inspired him to explore and collect specimens throughout the journey. For Darwin, these experiences, such as topographical changes he recorded after an earthquake in Chile, confirmed Lyell's theories.

Lyell's theories were grounded in the earth; he considered similar changes to the plants and creatures inhabiting earth unthinkable. Though he recognized that new species are created, he posited no ideas on how they came to be. Darwin, though, used his experiences from the *Beagle* to develop his theories of natural selection. As the earth changed gradually, so, too, did life forms change gradually over time. These theories explained not only the origin of species, but also the extinction. The processes of speciation and extinction should be so gradual as to be unnoticeable, yet scientists were able to observe the extinction of the great auk, which humans hunted into extinction.

(Great Auk: Photo courtesy of Wikipedia.org)

Darwin became familiar with humans' role in extinction during his *Beagle* travels, where he learned of the danger humans posed to a number of species.

He notes in *On the Origin of Species* that "through man's agency," animals become rare before they become extinct. If human interference with species does not constitute a special outside nature event, then possibly Cuvier was right and humankind is the catastrophe causing species extinction.

Key Take-Aways

- Charles Lyell put forth the idea that the earth forms and changes slowly, over many millions of years, both in the past and in the present.

- Charles Darwin applied Lyell's theories of the earth to its life forms, theorizing that they, too, change over long periods.

- Speciation and extinction should both take many generations, but extinction is happening faster due to the interference of humans.

CHAPTER 4

Summary

Gola del Bottaccione is a gorge in Italy. Here, geologist Walter Alvarez discovered the first evidence of the asteroid that ended the Cretaceous period and wiped out three-quarters of life on earth.

(Foraminifera: Photo courtesy of Wikipedia.org)

Foraminifera, forams for short, are miniscule sea creatures surrounded by shells called "tests." The shells fall to the ocean floor once their inhabitant dies, and each has a shape distinctive to its species. The abundance of these fossils makes them ideal for dating layers of rock. The limestone of Gola del Bottaccione presents a mystery. One layer contains forams of numerous species, all larger than average (still only the size of a grain of sand). The layer of clay above it has none, while the layer of limestone above the clay contains the miniscule forams of only a few species. The lowest level, with its many species of foram, suggests they died suddenly, at more or less the same time. Alvarez determined to study that strip of limestone to discover how much time it represented. He enlisted his father, a Nobel Prize-winning physicist, to help. Luis Alvarez theorized that testing the clay for iridium, a substance found in meteors and which constantly showers the earth as dust, would help date the three layers. The level of iridium in the clay layer was unbelievably high. Samples from Denmark and New Zealand, also from the late Cretaceous period, were tested for comparison and they, too, had off-the-chart levels of iridium.

Father and son theorized that "an

asteroid six miles wide" struck the earth with the force of "a hundred million megatons of TNT," causing mass extinction. The publication of their theory in 1980 generated huge interest within and without the science community, with one exception. Paleontologists felt they lacked basic understanding of paleontology, which was grounded in the theories of Lyell that say that extinction is a slow process and never instigated by cataclysmic events. Despite the dismissal of the paleontology community, corroborating evidence continued to accumulate. The clincher was the 100-mile wide crater in the Yucatán Peninsula, from which core samples revealed a layer of glass at the cretaceous-tertiary boundary period. This discovery ended the debate for most scientists. Eleven years after it was first posited, the theory that a meteor killed the dinosaurs was accepted.

Every living being today is descended from a creature that survived that impact. However, Darwin's theory that only the strongest survive does not apply here, as adaptation plays no role in surviving a catastrophic event. Traits that may have been advantageous for millions of years may suddenly become lethal.

Key Take-Aways

> • Fossil layers in limestone reveal that a catastrophic event 65 million years ago wiped out the vast majority of life on the planet. Testing reveals that event was a meteor impact that left a crater 100 miles wide.
>
> • Survival of the fittest only plays a role in species' survival during "normal" times, not during cataclysmic events.

CHAPTER 5

Summary

Experiments on perception show that preconceived ideas dramatically influence one's perception. Incongruous elements are ignored or explained away for as long as possible. Recognition inspires reactions similar to panic. This same mindset holds true in scientific disciplines, where "novelty emerges only with difficulty" until someone proclaims things may not be the way they've always been supposed to be. Then, the paradigm shifts, such as when naturalists had to admit all those strange bones couldn't be ignored, and Cuvier's theories were accepted. Then, as more and more bones were unearthed, the idea that catastrophe after catastrophe caused them no longer fit, so the theories of Lyell and Darwin were accepted. The uniformitarian view remained until the Alvarez work proved catastrophes did happen. The current paradigm is a combination of Cuvier and Darwin: "long periods of boredom interrupted occasionally by panic." The panic moments are rare but powerful.

Dob's Linn is a cliff in Scotland with

vertical, rather than horizontal, striations in the rock. These rocks date back 445 million years, to the Ordovician period. The continents of Africa, South America, Australia, and Antarctica were joined in a single land mass called Gondwana, while Scotland was in the southern hemisphere of Avalonia, which lay at the bottom of an ocean called Iapetus. The Ordovician period followed the Cambrian period. Both periods saw an explosion of new life forms, mainly water-based. Plants first appeared around the middle of the Ordovician. Toward the end, the oceans emptied, and approximately 85 percent of marine species died in what is now considered the first of the Big Five extinctions.

(Graptolite Fossils: Photo courtesy of Geograph.org.uk)

Graptolites were marine organisms that were nearly eradicated during the Ordovician event. To the naked eye, their fossils look like scratches. With the aid of a hand lens, the species' shapes appear: feather, lyre, fern fronds. They lived in colonies, attached to one another, and each fossil represents a community. Their presence is a significant aid in identifying layers of rock. The darker shale striations of Dob's Linn reveal numerous specimens, while the paler sections of rock are barren. This indicates abrupt changes where the sea floor became uninhabitable. The current theory of what wiped out life during the late Ordovician is glaciation. CO_2 levels dropped, temperatures fell, Gondwana froze, and sea levels plummeted. No one knows what caused these changes, but one theory is that the earth's new plants drew carbon dioxide out of the air, starting the whole glaciation process.

The opposite caused the end-Permian extinction. A massive, inexplicable release of carbon into the air caused temperatures to soar, and by the time it ended, 90 percent of species were dead. The cause of both the carbon release and massive loss of life are unknown. Once again, organisms that had evolved to thrive in their environment were faced with sudden, catastrophic change for which they were unprepared.

One day, the evidence of human civilization will be compressed to the width of a piece of paper within layers of stone. Monuments, art, climate change, and nuclear fallout all will be compressed and offer evidence of the age of humans. Nobel Prize-winning chemist Paul Crutzen suggested naming this age Anthropocene to reflect the many "geologic-scale changes" for which people are responsible. These actions have ensured the mark of humans will be reflected in the earth's stratification for millennia after their extinction. The case for naming this age Anthropocene is currently under consideration.

Key Take-Aways

- Earth-changing cataclysm that results in mass extinction may happen in a number of ways, not only via meteor impact. In the past, drastic climate change has caused two episodes of mass extinction.

- Geologists are considering calling the current age Anthropocene to reflect the huge influence humankind has on the earth.

CHAPTER 6

Summary

Castello Aragonese is a small Italian island, produced by shifting tectonic plates. The shifting occasionally causes volcanic eruptions. Through vents on the ocean floor, gas bubbles, which are nearly 100 percent carbon dioxide, regularly rise to the surface. Carbon dioxide forms an acid when it dissolves in water.

Due to burning fossil fuels, the amount of carbon dioxide in the air has been rising steadily since the beginning of the industrial revolution. Each year, humans add another nine billion tons (give or take), with the amount increasing by an average of 6 percent annually. The levels of carbon dioxide in the air today are higher than at any point in the last 800,000 years, and possibly several million years. If levels continue to rise at their current rate, the global temperature will increase between 3.5 and 7 degrees Fahrenheit, triggering world-altering events, such as melting the polar icecaps and destroying the remaining glaciers.

The other half of the CO_2 story involves the oceans, which comprise 70 percent of earth's surface. Gases from the atmosphere are absorbed by the ocean; gases dissolved in the ocean are released into the atmosphere. This exchange becomes uneven when the atmosphere's composition is changed, as humans have done with excessive fossil fuel consumption. Today, more CO_2 enters the water than leaves it. Each day, every American dumps 7 lbs. of carbon into the sea. This has changed the pH levels of the oceans globally, which are now 30 percent more acidic than they were in 1800. If the trend continues, by century's end, they'll be 150 percent more acidic. Thanks to the carbon dioxide-spewing vents, the waters surrounding Castello Aragonese provide scientists a preview of the oceans' future. Shells of the specimens removed from the water near the vents have been eaten away, discolored, and warped. Even more alarming, a third of the life forms found in the vent-free zone are missing, unable to survive in these highly acidic waters.

Ocean acidification played a large part in at least two of the Big Five extinctions, and it was possibly a major factor in the end-Cretaceous event that wiped out the dinosaurs. There is also evidence that it played a role in two other mass extinctions of marine life. If humans were pumping carbon dioxide out at a slower rate, it wouldn't have the same impact. The rapidity with which humans are burning through carbon that has been buried for tens of millions of years is unprecedented and likely cataclysmic.

Key Take-Aways

- Carbon dioxide is poisoning not only earth's atmosphere but its oceans as well.

- Areas of the ocean with high levels of carbon dioxide show a significant reduction in marine life, as well as damage to remaining organisms.

CHAPTER 7

Summary

One Tree Island is at the southernmost tip of the Great Barrier Reef. Formed entirely of coral rubble, the island was created 4,000 years ago during a terrible storm, and it continues to take shape today.

Despite never having seen one, Lyell theorized that coral reefs grew from the rims of extinct underwater volcanoes. Darwin's observations led to the correct theory that, if the island were to sink away over time, the atoll would remain.

(Photo courtesy of Wikimedia.org)

Coral reefs are comprised of individual corals called polyps that produce shells through calcification (extracting ions of calcium and carbonate from the water). Over generations, they work together to build the colony's exoskeleton. A reef is comprised of billions of polyps from hundreds of species. Thousands, if not millions, of species rely on coral reefs. Though it has been this way throughout many geologic epochs, scientists believe the ecosystem will become extinct by the end this century.

Corals grow to the height of the water level at low tide, and then spread out laterally, creating an expansive, flat reef. Coral reefs are found near Belize, the tropical Pacific, in the Indian Ocean and Red Sea, and in the Caribbean. However, the first evidence of the damage wrought by CO2 came from Biosphere 2 in Arizona. The atmosphere of Biosphere 2 became horribly polluted by CO2 when decomposition (which burns oxygen) overran photosynthesis (which creates oxygen). CO2 levels rose to six times the levels found outside the dome, wreaking havoc on the artificial ocean, wiping out most of the fish, and devastating its corals. Attempts to alter the water chemistry failed.

An experiment was begun to measure the saturation state (a measure of the calcium and carbonate ions) and its effect on corals. An abundance of CO_2 in water lowers the saturation state. Lowered states lower the growth rate of corals, which grow best at saturation state 5 and slow until they stop building at a level of 2. In today's oceans, saturation states above 4 are exceedingly rare. If current emission levels continue, there will no regions above 3.5 by 2060, and by 2100, none will be above 3. Reefs contain a staggering number of life forms and must be constantly growing just to remain the same. They are vulnerable to more than acidification. Threats include overfishing, agricultural runoff, deforestation, dynamite fishing, and climate change. If corals go, the entire ecosystem will follow.

Key Take-Aways

- Coral reefs form an intricate ecosystem that supports thousands of species.

- Increased levels of CO2 impact the growth of reefs, endangering the structure and all of its inhabitants.

CHAPTER 8

Summary

Popular opinion is that global warming poses the biggest threat to cold-loving species. However, global warming is expected to have an even greater impact on the tropics, in part because most species live in the tropics. Pick a spot at the north pole and head south (the only possible direction). A person will have to travel well over 2,000 miles before reaching the treeline of the boreal forests.

(Boreal Forest: Photo courtesy of Wikipedia.org)

Canada's boreal forest covers nearly a

billion acres, yet contains only 20 or so tree species. Further south, tree diversity picks up, and it continues to increase as one heads south through Central America to South America. Entering Peru, a person will find 1,035 tree species, 50 times as in the entire billion acres of Canada's boreal forest. This same diversity is found in every other life form (except aphids), which scientists call the latitudinal diversity gradient (LDG), meaning the lower latitudes have the richest biological variety. There are a number of theories to explain this discrepancy.

The evolutionary clock ticks faster in the tropics, allowing organisms to produce more generations, increasing the frequency of genetic mutations, which increases the likelihood of new species.

Species in the tropics need the relative stability of temperature found in the tropics. Populations become more isolated, sparking speciation.

The tropics are old, millions of years older than the forests of Canada, which was covered by mile-thick ice only 20,000 years ago. This means there has been far more time for diversity to accumulate in the tropics.

In Peru, variants in elevation of as little as 800 feet produce temperature variations of 2 to 3 degrees. Trees found at one elevation are often missing entirely from elevations only 800 feet above or below, suggesting their susceptibility to temperature. Studies of the trees migratory patterns reveal the effects of global warming. Trees "migrate" by dispersing seeds. After four years, researchers discovered trees were already migrating at an average of 8 feet per year. This average includes trees that spread up 800 feet while others stayed, for the most part, motionless.

Earth's temperatures have always fluctuated, warming and cooling repeatedly over many epochs and life forms adapted by migrating. The difference with the current warming phase is the rate at which it's happening, which is at least 10 times faster than the end of the last glaciation and all preceding glaciations. This means the world's life forms will have to migrate 10 times as quickly to keep pace. The question is: how many species will manage to keep up?

Key Take-Aways

- Global warming will have a large impact on tropical regions due to the greater abundance of species found there.

- The rate of warming is significantly greater than in previous warming periods. This will present a challenge to all life forms, whose response must keep up to ensure survival of the species.

CHAPTER 9

Summary

The Amazon is filled with "islands" with clinical-sounding names: Reserve 1202, Reserve 1112, etc. They are part of an experiment called the Biological Dynamics of Forest Fragments Project (BDFFP). Scientists from numerous disciplines study the BDFFP.

(BDFFP: Photo courtesy of Flickr.com)

The experiment began as a collaboration between ranchers and conservationists after Brazil's government incentivized ranchers to settle uninhabited areas of the rainforest. Concurrently, a Brazilian law required

Amazon landholders to leave half of the forest on their property unscathed. Biologist Tom Lovejoy had the idea to convince the ranchers to allow scientists to choose the 50 percent to use for an immense experiment. Brazilian officials embraced Lovejoy's plan, which has been in operation for over 30 years. The BDFFP is considered "the most important ecological experiment ever done."

Approximately 50 million square miles of earth's land are ice-free, more than half of which has been directly transformed by people. About 60 percent of the remaining land is natural forest; the remaining land is high mountain, tundra, or desert. Most scientists label only 11 million square miles of the earth "wildlands," and even these display evidence of humanity. BDFFP's reserves, with their unnatural outlines, reflect the rest of the planet.

Ornithologists banded nearly 25,000 birds before the forest was fragmented into reserves, and again afterward, for comparison. During the first year after the surrounding forest was cut down, the number of birds in the reserves increased as birds from the deforested area migrated. Over time, though, the number and variety of species decreased steadily. This degradation of diversity was reflected in other life forms, as well.

Natural islands demonstrate a lack of species diversity. One reason is that the area is inadequate to support certain species, both in size and resources. Another reason is that smaller populations are more vulnerable to chance. With the island's seclusion, the likelihood other creatures will recolonize is lower. This same phenomenon is true for "unnatural" islands like the reserves in the BDFFP. The absence of recolonization may cause local extinctions to become global.

The variety of life in the tropics is immense. Increased species diversity brings "low population density." This increases speciation but also vulnerability, as their isolation makes them more susceptible to extinction. Conservative estimates place the number of species populating the rainforest at two million. Assuming 1 percent of the forests are felled annually (another conservative estimate); the number of species going extinct each year is 5,000, or 14 per day.

Key Take-Aways

- The Biological Dynamics of Forest Fragments Project (BDFFP) is a huge experiment on life in the Amazon rainforest and the consequences of deforestation.

- Tropical areas have a great variety of life, but they are also highly vulnerable to extinction.

CHAPTER 10

Summary

In 2007, researchers in New York went to conduct a routine census on hibernating bats, but they were confronted with a cave full of dead bats, victims of a then-unknown fungus called psychrophile that arrived in America via Europe. The following winter, it spread to four states, and the winter after that to five more. It continues spreading and killing bats, over six million in North America to date.

Without the influence of people, ocean and land barriers would limit species profligation. Ocean barriers explain why continents with similar resources and landscapes contain vastly different flora and fauna. Land barriers keep the marine life of one ocean distinct from another. A distinguishing characteristic of Anthropocene is its interference with the "principles of geographic distribution." The process began with early human migration and has increased so dramatically that an estimated 10,000 species migrate every 24 hours, just in ballast water.

There are two possible scenarios when an organism migrates to a new area. The first is that nothing happens. There are many reasons the organism may not survive migration: the new climate is unsuitable, resources are too low, and it may be eaten. This is the fate of the vast majority. The second scenario is that the newcomer not only survives but also creates new generations in a process called "establishment." The danger comes when established species spread. Estimates are that approximately 10 percent of migrating species will become established, and 10 percent of established species will spread.

There are a number of reasons a species may spread. One is that its natural predators were left behind. A second scenario is that the flora and fauna of a new region aren't prepared for this new predator, which, without competition, may ravage the landscape. These invading animals are much like humans: succeeding at the expense of other species. When pathogens enter in the form of a fungus, bacteria, or virus, native organisms have evolved no defense against them, and populations are quickly wiped out.

Introduced species are so common that one would likely see numerous examples simply by looking out a window. There are databases set up around the world to track invasive species. The shuffling of species to new continents is akin to creating one giant continent that some biologists call the New Pangaea. This has greatly increased local diversity but reduced global diversity.

Key Take-Aways

> • Bats in North America are being wiped out by a fungus that humans brought from Europe. European bats are not susceptible to this fungus that has killed over six million in North America.
>
> • Invasive species may prove highly dangerous to native flora and fauna. Humans are generally responsible for the introduction of invasive species.

CHAPTER 11

Summary

In the nineteenth century, the Sumatran rhino was so common that it was considered an agricultural pest. However, as the forests of southeast Asia were felled, so was the rhino's habitat, and its population substantially reduced to just a few hundred by the 1980s. To ward off extinction, conservationists devised a captive breeding program. Of the 40 rhinos captured for this purpose, 12 died quickly. After 10 years, not one produced an offspring. A pregnancy was finally gained at the Cincinnati zoo, but it ended in miscarriage, as did four subsequent pregnancies. Finally, the sixth pregnancy resulted in the birth of a male, followed by a female and then another male. One male fathered another; these four are the only Sumatran rhinos born in the past 30 years.

(Sumatran Rhinoceros: Photo courtesy of Wikipedia.org)

The plight of the Sumatran rhino is reflected in other rhino species as well as most large mammal species, which are increasingly vulnerable due to poaching, hunting, and loss of habitat. Full-grown "oversized" mammals have no natural predators. From the time of the dinosaurs, jumbo-sized animals have been found throughout the world. These mammoth beasts disappeared in a relatively short amount of time from "over half the land surface of the globe." Why is debated, but the two prevailing theories are humans and climate change.

Arguments that humans are to blame have more traction, due to evidence that the mass extinctions took place in "pulses" and moved from continent to continent: Australia was hit 40,000 years ago, North and South America 25,000 years ago, Madagascar and New Zealand less than 1,000 years ago. This sequence aligns to human migration, as do archeological finds. Proponents of the climate change theory discount these happenings as correlation, not causation, and question how humans could have wiped out such large animals. Adherents of the human theory point out that the gestation period of large mammals is significant, between one and two years. Additionally, they give birth to single offspring. With no natural predators, their low reproduction doesn't present a problem. When an unnatural predator enters, though, this presents further evidence that "survival of the fittest" doesn't apply when catastrophe strikes. Further, the arrival of humans didn't immediately wipe out these species; it took hundreds of years. Humans continually hunting mammoth creatures with low reproduction rates over hundreds of years would easily lead to extinction. In "earth" time, this would seem instant. In "human" time, the animals' decline would hardly be noticeable.

Key Take-Aways

• Large mammals are endangered due to low birth rates and long gestation periods coupled with human interference: hunting, poaching, and habitat destruction.

• The first mass extinction of large mammals also occurred after human migration to their lands.

CHAPTER 12

Summary

Originally discovered in Germany, Neanderthal bones and tools have been found all over Europe and the Middle East. Before vanishing 30,000 years ago, Neanderthals lived in Europe for at least 100,000 years during a time of intense cold. After numerous theories as to why they vanished, it's generally believed humans were their downfall, after they arrived in Europe 40,000 years ago. Because humans had sex with Neanderthals before wiping them out, most people today are up to 4 percent Neanderthal.

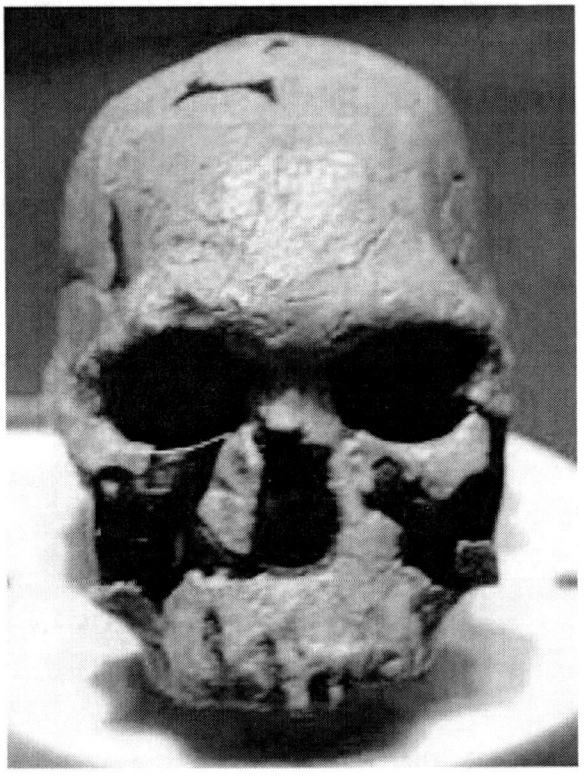

(Neanderthal: Photo courtesy of Flickr.com)

Svante Pääbo is a paleogeneticist working on sequencing the entire Neanderthal genome, which he will compare with the human genome to discover where the two species diverge. The process is difficult, to put it mildly, as DNA begins to disintegrate immediately upon death. The added complication of the discovery of the

Neanderthal bones led to heated debate among anti-evolutionaries, who insisted the bones belonged to an ordinary human. Numerous reasons were proposed to explain the misshapen bones, including rickets and too much time spent on a horse. As more bones were discovered, even evolutionists had trouble explaining the differences, such as larger skulls, which were at odds with then-current theories of small-brained apes leading progressively "to big-brained Victorians."

Early scientists dismissed Neanderthal as stupid, hairy, hunched, and brutish. In the 1950s, anatomists re-appraised the bones, pronouncing they likely walked upright. Further evidence suggests intelligence and an appearance not wholly dissimilar from modern humans.

The genome mapping proves that Neanderthals shared DNA with Europeans and Asians but not Africans. The explanation for this unexpected result is that, before wiping out Neanderthal, humans had sex with them, and those offspring helped populate Europe, Asia, and the New World. All non-Africans carry between 1 and 4 percent Neanderthal DNA.

Studies of Neanderthal bones reveal injuries and ailments that indicate a lack of projectile weapons (they'd have needed to attack prey from close quarters) and a sense of community wherein they'd care for a sick or injured member. The confinement of their bones to Europe and Asia suggests they did not cross borders such as open water, whereas humans did. Was it curiosity that led humans to cross dangerous open waters with no idea where those waters led? Will this thirst to explore be reflected when the comparison of the genomes of human and Neanderthal is complete?

Key Take-Aways

- Neanderthals populated Europe and Asia for over 100,000 years. Ten years after the arrival of humans, they were gone.

- Is there a genetic reason people explored their way into new territory? Is there a gene that makes humans "human?"

CHAPTER 13

Summary

At the Frozen Zoo, vials hold the cell lines of nearly 1,000 species, frozen in pools of liquid nitrogen. Around the world, a number of other institutions have assembled their own "chilled menageries." Currently, the majority of these species are endangered, not extinct, but that ratio is dwindling. Is there any way the tide can be turned? Humans have proven repeatedly their willingness to act to save endangered species and preserve earth's flora and fauna. Would the better course be to focus on what can be done and what is being done to save endangered species, rather than engage in gloomy prognostication? In a different kind of book, the answer would be yes.

Throughout the previous chapters' details of extinctions and near-extinctions, there has been a unifying theme: tracing the pattern of the Sixth Extinction. Earth's history makes clear that no matter how resilient a species, each has its breaking point. The cause of previous mass extinctions is varied: glaciation, global warming, asteroid. The one similarity in the Big Five is rate of change. Species that cannot adapt at the same rate expire, and the instigator of rapid change does not factor into the equation. People change the makeup of their world, and they have been doing so since they first appeared.

What will happen to the humans? Possibly, due to their disruption of "earth's biological and geochemical systems," people, too, will fall. History has proven that even dominant species are susceptible to cataclysmic change. A second possibility is that people will devise a way to survive, such as reengineering earth's atmosphere or heading to outer space. Right now, people are deciding what will survive and what will be lost forever. Long after humanity and all its contributions have disappeared forever, the Sixth Extinction will "continue to determine the course of life" on earth.

Key Take-Aways

- The Sixth Extinction is underway.

- Humanity is the cause of this newest extinction event.

PUTTING IT TOGETHER

Elizabeth Kolbert believes that, not only is earth in the middle of a humanity-caused Sixth Extinction, but that humans have a say in which species will go and which will remain. To support her theories, she offers readers detailed accounts of previous mass extinctions and examples of current species that are severely endangered.

The unifying theme of previous mass extinctions is rapid change; in the current mass extinction, humans are the origin of that change. In this context, "rapid" refers to geological time, in which thousands of years constitute a blink of the eye. Species unable to adapt quickly enough are not likely to survive these cataclysmic events. In these instances, dominant species are every bit as vulnerable as lesser organisms, as they've not evolved to withstand such catastrophes. In fact, the traits that make them dominant may contribute to their undoing as species.

Kolbert does an excellent job detailing the countless ways humans have instigated catastrophic change and how those changes impacted a region's native flora and fauna shortly after the arrival of migrating humans. Even those who deny climate change will have difficulty reading the histories presented in *The Sixth Extinction* and maintaining their stance that humans couldn't be responsible for the destruction that's followed in their wake for tens of thousands of years. Her work may even present a challenge to creationists' ability to continue their denials of evolution when confronted with the mountain of evidence presented by scientists who represent not only many disciplines but also four centuries of thought and research. Naturalists, biologists, geologists, paleontologists; if you can put an "ist" at the end of it, the research and work of that group seems to be represented in Kolbert's book, and to great effect.

What may be Kolbert's most intriguing theory is the idea that humans' cognizance of what is happening around the globe, their ability to reason, places them in the unique position of determining the fates of endangered species, and the fate of humanity, as well.

ABOUT THE BOOK'S AUTHOR

A long-time staff writer for *The New Yorker*, journalist Elizabeth Kolbert has won numerous awards for her work, including the National Magazine Award for Public Interest, the Lannan Literary Fellowship, National Academies Communication Award, 16th Annual Heinz Award (with special focus on global change), National Magazine Award for Commentary, and Guggenheim Foundation Fellowship in Science Writing.

Prior to publishing *The Sixth Extinction*, Kolbert published two other books: *The Prophet of Love*, which focuses on New York's public figures, and *Field Notes from a Catastrophe*, which tackles global warming. She has also edited and contributed to a number of other published works dealing with climatology. Her stories and articles have appeared in numerous publications, including *The New York Times Magazine*, *Vogue*, and *Mother Jones*.

After studying literature at Yale University, Elizabeth Kolbert transferred to Germany's Universität Hamburg upon winning a Fulbright Scholarship in 1983. While in Germany, she began working for

The New York Times as a stringer before moving on to the Metro desk, serving as the Albany bureau chief, and finally to writing the Metro Matters column. She joined *The New Yorker* in 1999, where she's published numerous pieces on climate change and politics, including profiles on Hillary Clinton and Rudy Giuliani. Her series on global warming, "The Climate of Man," appeared in *The New Yorker* in the spring of 2005 and won the American Association for the Advancement of Science's magazine award.

Elizabeth currently resides in Massachusetts with her husband and three sons, but she has traveled around the world to develop her understanding of global warming and climate change. She's visited caves in North America, rainforests in Peru, and coral reefs in Australia, to name just a few of the sites profiled in *The Sixth Extinction*, proving she's always willing to get down and dirty in the fight against extinction.

Readers Who Enjoyed This Ebook Might Also Enjoy...

Elizabeth Kolbert's *The Sixth Extinction* (the full book)

Eric Schmidt & Jared Cohen's *The New Digital Age*

Svante Paabo's *Neanderthal Man*

Save Time Summaries' *Grain Brain: The Surprising Truth about Wheat, Carbs, and Sugar (Your Brain's Silent Killers) by David Perlmutter -- Summary, Review & Analysis*

#

CPSIA information can be obtained at www.ICGtesting.com
Printed in the USA
LVOW05s1802230314

378579LV00016B/690/P